The Girl
and the
Birthday
Problem

by Dr Boaz

**Illustrated by
Chang Yit Wah**

**Includes a Guide to
Problem-solving by
Yeap Ban Har**

NEW JERSEY · LONDON · SINGAPORE · BEIJING · SHANGHAI · HONG KONG · TAIPEI · CHENNAI · TOKYO

Published by
WS Education, an imprint of
World Scientific Publishing Co. Pte. Ltd.
5 Toh Tuck Link, Singapore 596224
USA office: 27 Warren Street, Suite 401-402, Hackensack, NJ 07601
UK office: 57 Shelton Street, Covent Garden, London WC2H 9HE

National Library Board, Singapore Cataloguing in Publication Data
Name(s): Boaz, Dr. | Chang, Yit Wah, illustrator.
Title: The girl and the birthday problem / by Dr Boaz ;
         illustrated by Chang Yit Wah.
Other Title(s): I'm a math star.
Description: Singapore : WS Education, [2022] |
         Includes a guide to problem-solving by Yeap Ban Har.
Identifier(s): ISBN 978-981-12-5055-2 (hardcover) |
         978-981-12-5111-5 (paperback) | 978-981-12-5056-9 (ebook for institutions) |
978-981-12-5057-6 (ebook for individuals)
Subject(s): LCSH: Mathematical recreations. | Mathematics.
Classification: DDC 793.74--dc23

British Library Cataloguing-in-Publication Data
A catalogue record for this book is available from the British Library.

For any available supplementary material, please visit
https://www.worldscientific.com/worldscibooks/10.1142/12675#t=suppl

Desk Editor: Daniele Lee
Design: Eliz Ong

Printed in Singapore

It was the first day of school.

School was fun!

The girl was excited.

Her classmates were fun to learn with.

Her teacher, Miss Cheryl,
was very interesting.

All too soon, it was time to go home.

Daddy and Mummy came to fetch her.

Miss Cheryl and Mr Daud, the principal,
waited with the girl and greeted her parents.

"Lovely to meet you, Mr and Mrs Lee."

"The pleasure is ours.
Thank you for teaching our
little girl, and for waiting
with her till we arrive."

"Your little girl told me
that you love puzzles,"
said Miss Cheryl cheekily.
"Would you like one now?"

"Certainly," Mummy said,
with a twinkle in her eye.

"OK. I'm going to ask
you to guess
my birthday.
With some hints,
of course."

Miss Cheryl wrote 10 dates on the whiteboard in the classroom.

"My birthday is one of these dates."

Miss Cheryl took a
piece of paper,
wrote a word on it,
and gave it to Daddy.

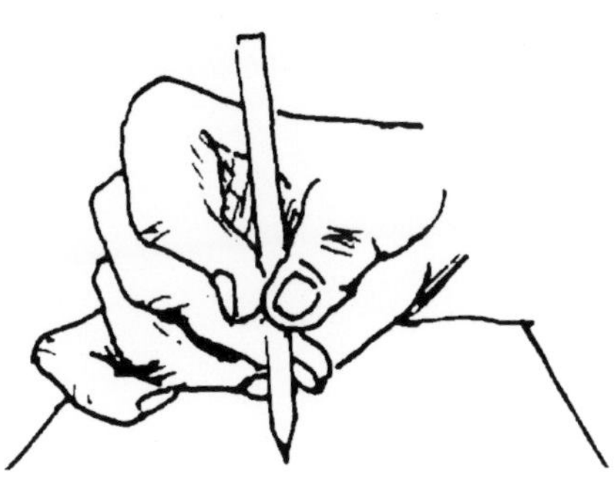

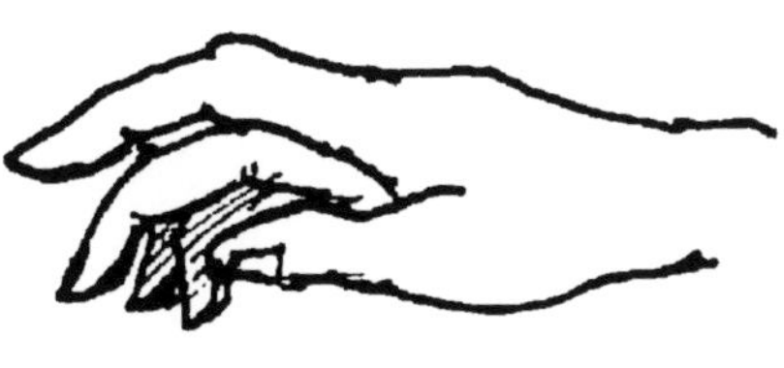

"It's the actual month, Mr Lee,"
she said cheerfully,
"and it's for your eyes only!"

Then, she took another
piece of paper, wrote a
number on it and gave it
to Mummy.

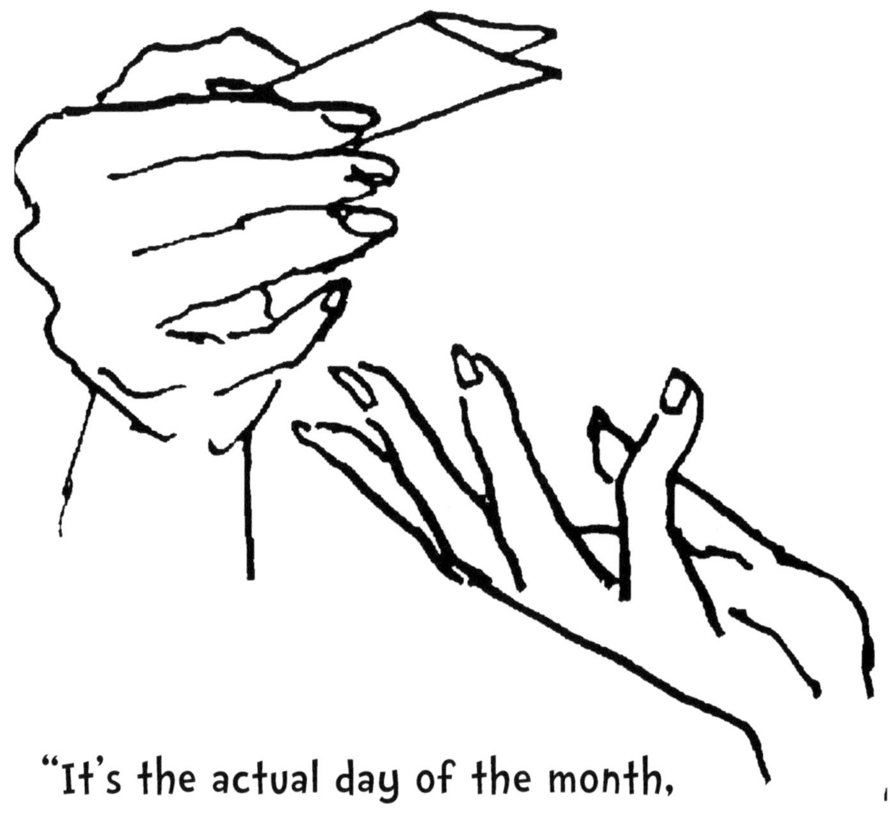

"It's the actual day of the month,
Mrs Lee," she said blithely,
"and it's for your eyes only!"

The girl looked on with great interest.
She waited to see what her parents
would say.

Daddy looked at the month on
his piece of paper

and he looked at the ten dates
on the whiteboard.

Then he said, "I don't know when Miss Cheryl's birthday is

but I know that Mummy doesn't know either!"

The girl looked intently at Daddy.
"Why would he say the second part?" she
whispered to herself.

Now, Mummy made some
notes on her piece of
paper

and said triumphantly,
"I know Miss Cheryl's birthday now!"

The girl was surprised but
before she could ask how,

Daddy pronounced,
"If Mummy know Miss Cheryl's birthday,
then I know it now too!"

The four adults laughed merrily.

The girl wanted to join in the fun.

"Can I guess Miss Cheryl's birthday as well?"

"Certainly, dear," Daddy said.

"Can I see either your piece of paper
or Mummy's piece?"

"But you won't need
either, my dear,"
Mummy smiled,

"if you had
listened carefully
to what we said."

The girl was lost for words ...

"Why don't you try stepping into our shoes to see why we said what we said," suggested Daddy.

"Now, that statement seems
strange," the girl thought.
"But Daddy and Mummy always
mean what they say.
OK then, let me try."

"But first, let me rewrite the ten dates in an easier way for me to see."

Jan	2		5			
Feb				7	11	
Mar	2	3				13
Apr		3	5	7		

And this is what she rewrote on one side of the whiteboard.

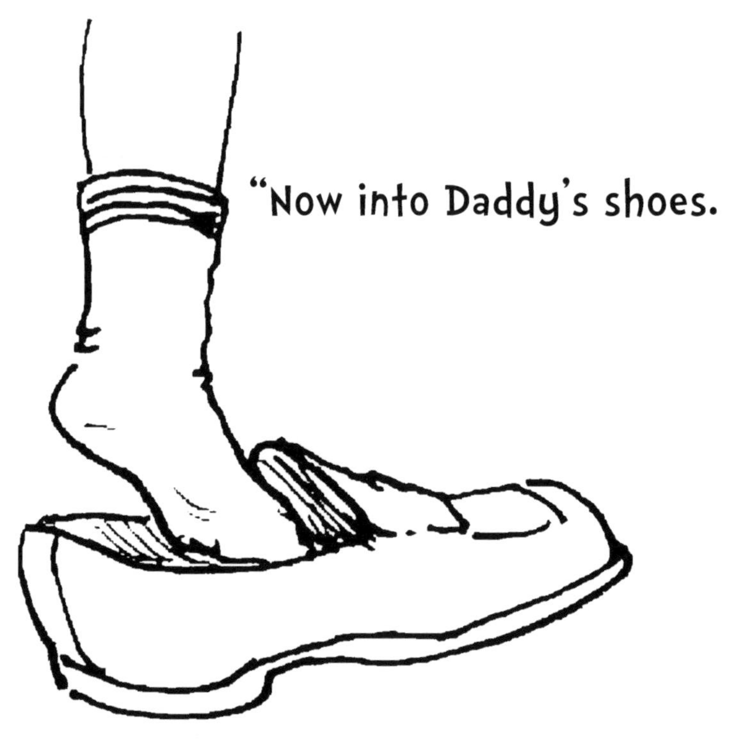

Daddy sees a month.

But which month?

I don't know, but it is easier if I

just take a firm possibility.

Suppose he sees April.

Jan	2		5			
Feb				7	11	
Mar	2	3				13
Apr		3	5	7		

He looks at the whiteboard
and then reasons that
Miss Cheryl's birthday in that
month falls on 3, 5 or 7.
But he doesn't know exactly.
So he says 'I don't know when
Miss Cheryl's birthday is.'

And then he adds, 'But I know
that Mummy also doesn't know.'

Why does he say that?"

The girl thought hard.

And then she thought even harder.

"I know! Daddy must have stepped into Mummy's shoes then!

Daddy may have thought that
Mummy sees the day 3, 5 or 7.

Jan	2		5			
Feb				7	11	
Mar	2	3				13
Apr		3	5	7		

Suppose she sees 3. She looks at the whiteboard and sees that there are two 3's, either 3 March or 3 April. So she won't know exactly when Miss Cheryl's birthday is.

Suppose she sees 5.

There is 5 January and there is 5 April.

So she won't know exactly

when Miss Cheryl's birthday is.

Suppose she sees 7.

There is 7 February and there is 7 April.

For all three possible numbers,

she won't know exactly

when Miss Cheryl's

birthday is.

That's why Daddy said,

'But I know that

Mummy also

doesn't know.'"

"What about the other months?"
the girl thought.
"I must patiently go through all of them."

Jan	2		5			
Feb				7	11	
Mar	2	3				13
Apr		3	5	7		

She checked that the days
2 and 5 in January
had duplicates in
2 March and 5 April.

Jan	2		5			
Feb				7	11	
Mar	2	3				13
Apr		3	5	7		

"So if Daddy saw January on his piece of paper, he would also be able to say, 'But I know that Mummy also doesn't know.'"

Now the girl checked for February
and was careful to notice
that the day 11 was not repeated
in any other month!

Jan	2		5				
Feb				7	**11**		
Mar	2	3					13
Apr		3	5	7			

"So Daddy would think,
'If Mummy had this day '11',
she would know
Miss Cheryl's birthday.'"

The girl did not notice the adults
looking at her with amusement.

"So Daddy wouldn't have been able to say,
'But I know that Mummy also doesn't know.'
And so Daddy couldn't have seen
'February' on his paper!"

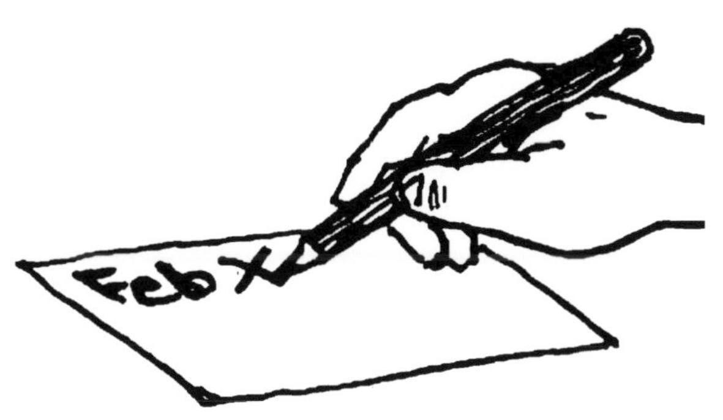

The girl was happy.
She said, "I now know that
Miss Cheryl's birthday
cannot be in February."

"That's wonderful, dear!"
Daddy beamed with pride.
"But, is that all?"

Jan	2		5			
Feb				7	✕	
Mar	2	3				✕
Apr		3	5	7		

The girl smiled and checked the
remaining month, March.
She quickly saw that the day 13
was not repeated in any other month.
"Miss Cheryl's birthday cannot be in March
either," she laughed.

The girl walked to the whiteboard
and put a long line across each of the
February and March rows.

"That's to make things even
clearer for the next step," she
said with a determined look on
her face.

"Now into Mummy's shoes...

...After Mummy heard
Daddy's words, she would
have crossed out February
and March, just like I did.
So, this is what Mummy
would have seen, and based
on that, she concluded,
'I know Miss Cheryl's
birthday now!'

Jan	2		5			
Apr		3	5	7		

The only numbers left are 2, 3, 5 and 7."

The girl stared and stared at the board.
"But how would I know what number she sees
in the paper Miss Cheryl gave her?"

The girl looked dejected.

"Be patient, little girl.
You are almost there,"
the principal comforted her.

"Perhaps, you
could go through
the days just as
you went through
the months ..."

The girl cheered up.
"I'm going to be
patient and calm,"
she said, feeling
determined.

"Now, if Mummy had the number 5 written on her paper, she wouldn't know which month to choose since 5 is present in both January and April!

Jan	2		5			
Apr		3	5	7		

Yes, I know that

Mummy doesn't have 5."

And she struck off the number 5
with a fierce vertical line!

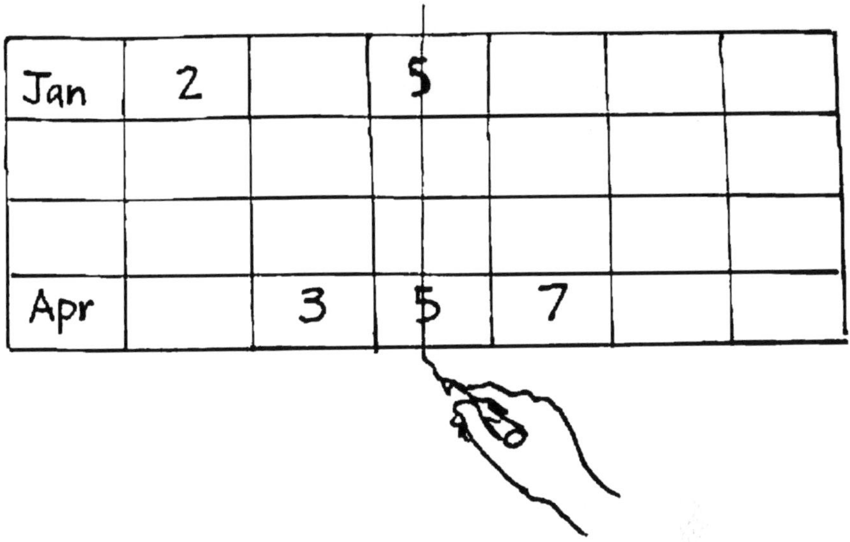

"That leaves 2, 3 or 7 and Mummy knew the
exact birthdate because she had 2, 3 or 7
written on her paper. It's 2 January, 3 April
or 7 April!"

"Bravo!" Mummy cheered.

With a playful grin,
the girl pretended
to put on some shoes.

"Whose shoes are you
putting on now, dear?"
Daddy asked.

"Why, yours,
of course!
At this point,
you could have
had the month
January or April
written on
your paper.

Let me think ...

If you had April written on your paper, you could not have said that you now know the birthday, because you would have had two options in that month: 3 April or 7 April. Since you said you knew the birthday, you must have had the month January, because that is the month with only one day option: 2 January!"

The girl paused, smiled...

... and triumphantly, she said,
"Happy birthday, Miss Cheryl
it's the 2nd of January today!"

And they all burst out laughing.

And the girl was happy.

The End

The *I'm a Maths Star!* Guide to Problem-solving

Yeap Ban Har

What can we do when faced with a challenging Mathematics problem such as the one in *The Girl and the Birthday Problem*? How can we prepare ourselves to be good at problem-solving?

Here are eight tips to keep in mind:

Tip #1:

Be like the girl: be calm when facing a challenging problem

- The girl did not panic. Neither did she give up. She was calm and did what she could at that moment. She rewrote the ten dates in a way that made sense to her.
- A heuristic is a way of solving problems that is not guaranteed to be the best or perfect, but is enough for us to make a small step forward.
- Use a heuristic when encountering a challenging problem. We can take small steps forward and **do what we can at that moment**.

Tip #2:

Be like the girl: solve parts of the problem

"Now into Daddy's shoes."

- The girl broke the problem down into smaller parts. She started by stepping into Daddy's shoes and later into Mummy's shoes. Finally, she went back to see things from Daddy's point of view.
- Solve a problem **part by part**. The solution to one part may often help us progress in solving a complex problem.

Tip #3:

Be like the girl: make a supposition

"Daddy sees a month. But which month? I don't know, but it is easier if I just take a firm possibility. Suppose he sees April."

- In this problem, the answer is one of the four months Miss Cheryl wrote down.
- By supposing the birthday was in April, even if it was not, the girl was able to understand the problem better and realised what she needed to do next.

- By making a supposition, or **making a guess, and then checking**, we can understand a problem better. Often, doing this will help us see the next step.

Tip #4:

Be like the girl: be systematic

"What about the other months?" the girl thought. "I must patiently go through all of them."

- After getting some insights if the answer was April, the girl worked systematically and applied her understanding to the other three months, one at a time.
- We will not miss out important possibilities when we are **systematic in our approach**.
- By being systematic, we will also strengthen our understanding of the problem by applying the same logic to the other cases. This will often open up doors to the final solution.

Tip #5:

Be like Daddy: monitor your thinking

- The girl gained important insight by considering April. She then used the same thinking for January and February. Daddy assured the girl that she was on the right track but questioned if she had missed anything out.
- Daddy personifies metacognition, our mind's ability to monitor itself. The principal too acts as the girl's metacognition.
- Metacognition alerts us if we are on the right track or otherwise.

Metacognition also helps us realise we are not yet done and prompts us to continue. Metacognition is important in becoming good problem solvers.
- We may start off like the girl— being on the right track. However, we must make sure we end up like Daddy and **check every single case.**
- And like the principal said, **we must push ahead** using what we already understood when things seem too hard. The solution will follow.

Tip #6:
Be like Mummy: enjoy figuring things out
"Your little girl told me that you love puzzles,"...

- When we **enjoy figuring things out**, we learn better.
- When we enjoy figuring things out, we are **training our mind** to become better at figuring things out.

Tip #7:
Be like the girl: be motivated

- We need to **want to be successful** to be successful. This is also true for solving Maths problems.

Tip #8:
Be like the girl: find joy in solving problems

- **Even if we don't arrive at a solution right away, it is fine.** The process itself makes us a better thinker.
- When we take the time to **celebrate solving problems**, we will gain motivation to solve further problems we encounter.

For parents and teachers

Pólya's Model of Problem-solving

We would suggest that any attempt at mathematical problem-solving requires a model to which the problem solver can refer, especially when one is unable to progress satisfactorily. A model helps in regulating a problem-solving attempt. We shall now describe the essential features of the problem-solving model proposed by George Pólya, an eminent Hungarian mathematician. The process can be summarised in the diagram below.

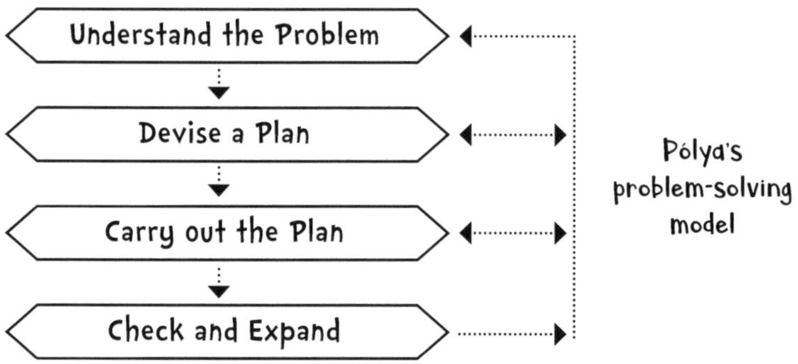

The model resembles a flowchart with four stages, Understand the Problem, Devise a Plan, Carry out the Plan, and Check and Expand, which are hierarchical but which allow back-flow. We shall explain the stages of the model and how the model works by applying it to our story of *The Girl and the Birthday Problem*.

Understand the Problem

Read aloud with the child. Pause to admire the nice illustrations. Take time to discuss and answer the child's questions, especially with regard to what the two main persons say and when they say them. The key parts of the problem to take time to understand would be:

- Daddy and Mummy know different bits of information — Month and Day
- Daddy and Mummy pick up more information each time the other talks
- The girl picks up more information each time Daddy or Mummy talks

It is common to see people staring blankly when they are stuck at a problem. It is important that something is done and the following describes how heuristics (meaning: serving to find out or discover, *Oxford English Dictionary*) can help start the problem-solving process.

Three useful heuristics to help understand the problem are to act it out, divide into cases and to eliminate possibilities (through contradiction). Consider the problem supposing Daddy sees the month of April. Then patiently act it out putting oneself into someone else's shoes. Greater clarity will result from this worthwhile use of time.

Devise a Plan / Carry out the Plan

A plan is vital to the success of any endeavour. It has been said, "He who fails to plan, plans to fail."

In the story, the girl's plan was to rewrite the 10 possible dates into a more manageable table. She also decided to act out what her Daddy and Mummy might have been thinking according to what they said. This resulted in a humorous stepping into different shoes and step by step (pun intended) elimination of wrong dates!

Check and Expand

It is good to look back at the solution and check it again. Check if Daddy holding January and Mummy holding 2 would indeed result in what each of them said.

We are done for this problem but not done with the problem-solving process. It is a key feature of the model that the solver should try to 'expand' the problem even though it is solved. By expanding, we mean one of the following:
- understand the underlying structure of the solution
- pose new problems

Indeed, if we understand the elimination of possibilities due to the existence or nonexistence of doubles, we could ask the child to pose a similar problem for his or her own birthday!

By thinking through the new problems, the child will reinforce his or her understanding of the original problem.

Scan this QR code to get additional practice on how to solve similar problems. Solutions are provided too.

Learn to solve difficult Maths problems using Maths heuristics, as identified by the Singapore Maths curriculum!

The *I'm a Maths Star!* series comprises challenging Maths puzzles, presented in story form. Puzzles are described and then solved, step-by-step, through an engaging storyline, and using various Maths heuristic techniques that are also taught as part of the world-renowned Singapore Maths curriculum. Through illustrations and an engaging storyline, children will learn Maths heuristics, and be inspired to persevere in understanding, representing, and solving fun and intriguing Maths problems!